C000156682

# About the book

There are 4, year 8 Mathematics papers & answers in this book. These are 2 sets of papers 1 (non-calculator) & 2 (calculator) written as practice papers for end of year 8 Mathematics Examinations in June 2020. Papers are mainly focusing on topics covered by most schools in year 8 mathematics syllabuses in The United Kingdom. However, you may still use this book as a practice for other syllabuses for 12 to 13 year olds.

All the questions in this book are written by the author and they are new questions written purely to help and experience the students to prepare and test themselves for the upcoming end of year mathematics exams.

Answers are included in this book. If you need to check your solutions, I advise you to ask your school mathematics teacher or your private mathematics tutor to mark your answers.

There are 2 sections to this book A & B. Each section contains 2 papers. The first paper of each section is a non-calculator paper & the second paper of each section is a calculator paper.

# Year 8 Mathematics Practice Papers

## (Year 8 Mock Exams)

### for 12 to 13 year olds

### 4 mock papers including answers

**By Dilan Wimalasena**

# Contents

# Section A

# Year 8

# Mathematics

# Practice Paper A1

# June 2020

Calculator is not allowed

**Time allowed**
**1 hour**
**Total 100 marks**

Write answers in the space provided

1. Work out the following and write the answers in their simplest form.

$i)\ \dfrac{8}{9} \times \dfrac{3}{4}$

(3 marks)

$ii)\ \dfrac{16}{21} \times \dfrac{14}{24}$

(3 marks)
(total 6 marks)

2. Venn diagram below shows two sets A & B.

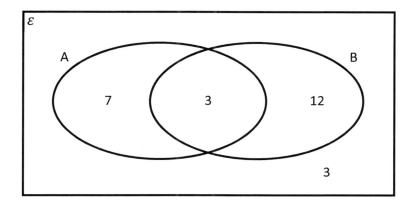

i) How many people are in sets A or B or both?

(3 marks)

ii) What is the probability of A?

(3 marks)
(total 6 marks)

3. John walked 12km in 3 hours. Calculate his average speed in $km/h$.

(3 marks)

4. Expand and simplify.

$i)\ 2(x + 2) + 3(x + 5) =$

(4 marks)

$ii)\ x(x + 2) + 3(x + 2) =$

(3 marks)
(total 7 marks)

5. A sequence has terms $2, 5, 8, ... ... ...$

i) Find an expression for the $n^{th}$ term.

(3 marks)

ii) Find $50^{th}$ term?

(3 marks)
(total 6 marks)

6. Solve the following equations.

$i)\ 2(3x - 1) + 5 = 21$

(4 marks)

$ii)\ 4x - 3 = 5(x - 2)$

(4 marks)
(total 8 marks)

7. Convert the following percentages into fractions.
i) 75%

(2 marks)

ii) 28%

(2 marks)

iii) 35%

(2 marks)

iv) 48%

(2 marks)
(total 8 marks)

8. John and Mary shared 24 sweets in ratio 3:5.
Work out each share?

(4 marks)

9. Work out the following.
i) $2^3$

(2 marks)

ii) $7^2$

(2 marks)

iii) $8^3$

(3 marks)

iv) $4^4$

(3 marks)
(total 10 marks)

10. Bearing of B from A is 060°. Distance from A to B is 8km. Using a scale of 1cm to 1km, construct the points A & B clearly showing the bearing in the space below.

(5 marks)

11. i) Convert 10kg into grams.

(2 marks)

ii) Convert 1500ml into litres.

(3 marks)
(total 5 marks)

12. Work out the value of $x$ in each diagram.

i)

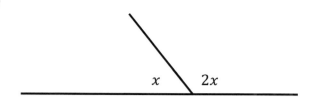

(3 marks)

ii)

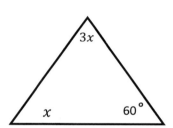

(3 marks)

iii)

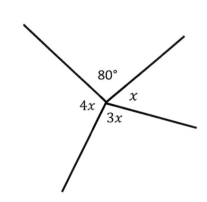

(4 marks)
(total 10 marks)

13. Factorise fully.
$i)\ 3x + 15 =$

(2 marks)

$ii)\ 4y - 10 =$

(2 marks)

iii) $x^2 + 7x =$

(2 marks)

$iv)\ y^2 - 8y =$

(2 marks)

$v)\ 3x^2 + 12x =$

(3 marks)

$vi)\ 10y^2 + 15y =$

(3 marks)

(total 14 marks)

14. Find highest common factor (HCF) and lowest common multiple (LCM) of
i) 48 & 64

(4 marks)

ii) 36 & 54

(4 marks)

(total 8 marks)

Total for paper: 100 marks

**End**

# Year 8

# Mathematics

# Practice Paper A2

# June 2020

Calculator is allowed

**Time allowed**
**1 hour**
**Total 100 marks**

Write answers in the space provided

1. John bought 8 bags of potatoes for £4.80. Calculate the cost of 10 bags of potatoes.

(4 marks)

2. Write following list of numbers in ascending order.

0.2010,     0.2101,     0.2111,     0.1021,     0.0210,     0.1002

(3 marks)

3. A circle has radius 3cm.
i) Calculate the circumference of the circle.

(2 marks)

ii) Calculate the area of the circle.

(3 marks)
(total 5 marks)

4. i) What is 30% of £150?

(3 marks)

ii) What is 45% of £200?

(3 marks)
(total 6 marks)

5. Write the following numbers as a multiple of their prime factors.
i) 100

ii) 120

iii) 300

(4 marks)
(total 10 marks)

6.

Area 27.3 $cm^2$

Length 6.5 cm

i) Calculate the width of the rectangle.

(4 marks)

ii) Hence, work out the perimeter of the rectangle.

(3 marks)
(total 7 marks)

7. Work out angles $a, b$ & $c$. (write reasons)

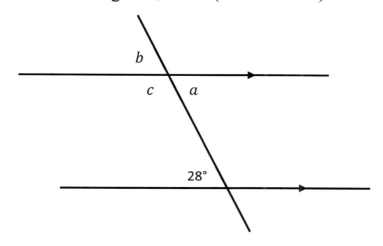

(9 marks)

8. Below is some data for ages of 25 students.

| Age ($x$) | Students ($f$) | |
|---|---|---|
| 12 | 3 | |
| 13 | 7 | |
| 14 | 8 | |
| 15 | 5 | |
| 16 | 2 | |

Work out the average age of students.

(6 marks)

9. Work out the following

$$3\frac{1}{4} - 2\frac{1}{3}$$

(5 marks)

10. A car is priced at £12,500. A 15% discount is offered.

i) Calculate the discount amount.

(3 marks)

ii) Calculate the price after the discount.

(2 marks)
(total 5 marks)

11. Calculate the following
i) $\sqrt{625}$

(2 marks)

ii) $\sqrt{961}$

(2 marks)

iii) $\sqrt{961} - \sqrt{625}$

(2 marks)
(total 6 marks)

12. Convert the following into metres.
i) 1.75km

(2 marks)

ii) 765cm

(2 marks)

iii) 0.25km

(3 marks)

iv) 82cm

(3 marks)
(total 10 marks)

13. Draw the net for the cube drawn below.

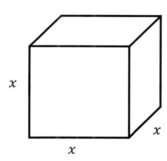

(4 marks)

14. $y = 3x + 1$

i) Complete the table below.

| $x$ | $-2$ | $-1$ | 0 | 1 | 2 | 3 |
|---|---|---|---|---|---|---|
| $y$ | | | | | | |

(4 marks)

ii) Plot the straight-line graph of $y = 3x + 1$.

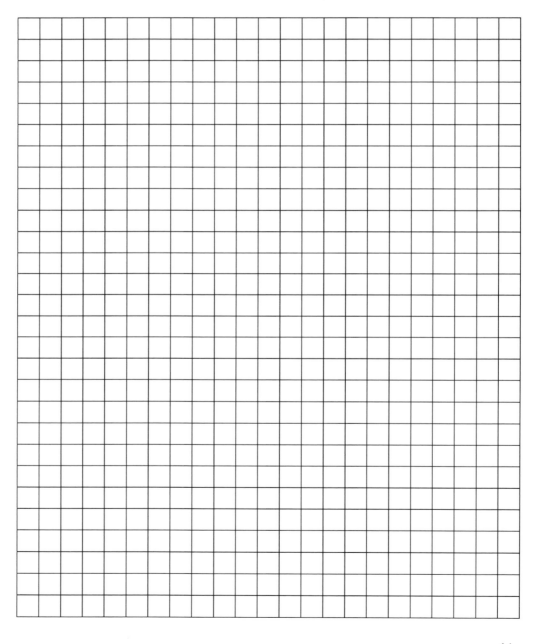

(4 marks)
(total 8 marks)

15. Work out the area of the shape drawn below.

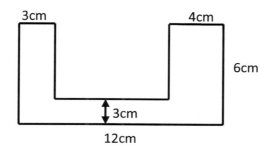

(6 marks)

16. Work out the following.
$i) -2 + (3 \times 4) =$

(3 marks)

$ii) (-30 \div 5) + (4 \times -2) =$

(3 marks)
(total 6 marks)

Total for paper: 100 marks

**End**

# Section B

# Year 8

# Mathematics

# Practice Paper B1

# June 2020

Calculator is not allowed

**Time allowed
1 hour
Total 100 marks**

Write answers in the space provided

1. Estimate the following.

$$\frac{99.5 \times 5.01}{6.99^2}$$

(5 marks)

2. Work out and simplify.

i) $3\frac{1}{3} \times 2\frac{2}{5}$

(4 marks)

ii) $2\frac{1}{3} \div 1\frac{1}{6}$

(4 marks)
(total 8 marks)

3. Work out.
i) $2.3 \times 1.4$

(3 marks)

ii) $4.12 \times 5.6$

(4 marks)
(total 7 marks)

4. Jenny has 70% chance of winning a game. She plays 50 games. How many games is she expected to win?

(4 marks)

5.

4cm

8.5cm

i) Calculate the area of rectangle.

(4 marks)

ii) Calculate the perimeter of rectangle.

(4 marks)
(total 8 marks)

6. Books cost £5 each and pens cost £2 each. Steve bought $x$ books and $y$ pens. Write down an algebraic expression for the price (P), he paid for his items.

(4 marks)

7. A sequence has terms

    4.3,     5,     5.7,     ……,     ……,     7.8,     ……

Work out the missing terms.

(6 marks)

8. The volume of the cuboid below is $60cm^3$. Work out the value $x$.

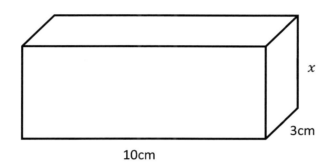

10cm

3cm

$x$

(5 marks)

9. Here are some scores by a cricket team in their last 7 matches.

    123,     125,     115,     128,     125,     131,     134

i) What is the mode?

(3 marks)

ii) What is the median?

(3 marks)
(total 6 marks)

10. Work out the value of angles $x$, $y$ & $z$. (write reasons for each case)

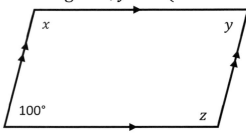

(12 marks)

11. Work out. (clearly show your steps)
i) $100 \div 8$

(3 marks)

ii) $216 \div 5$

(4 marks)
(total 7 marks)

12. Simplify fully

i) $x^3 \times x^4 =$

(2 marks)

ii) $x^8 \div x^3 =$

(2 marks)

iii) $\dfrac{x^{10}}{x^7} =$

(2 marks)

iv) $\dfrac{x^5 \times x^8}{x^2 \times x^4} =$

(5 marks)
(total 11 marks)

13. Work out the value of angles $a$ & $b$.

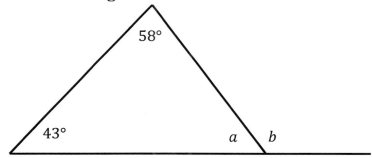

(6 marks)

14. Solve the following equations.

$i)$ $5(x + 3) + 2(x - 1) = 41$

(4 marks)

$ii)$ $2a - 7 = 9 - 6a$

(4 marks)
(total 8 marks)

15. Write 42 as a product of its prime factors.

(3 marks)

Total for paper: 100 marks

**End**

# Year 8

# Mathematics

## Practice Paper B2

## June 2020

Calculator is allowed

**Time allowed
1 hour
Total 100 marks**

Write answers in the space provided

1. A bag of potatoes costs £1.54 and a bottle of milk costs £1.19. Mary bought 2 bags of potatoes and 2 bottles of milk. Calculate her total bill.

(5 marks)

2. Find highest common factor (HCF) and lowest common multiple (LCM) of
i) 48 & 80

(5 marks)

ii) 72 & 90

(5 marks)
(total 10 marks)

3. Joanna drove her car at an average speed of 40km/h for 2.5 hours. Work out the distance, she has travelled.

(4 marks)

4. Solve the following equations.
(leave your answers as fractions where appropriate)

$i)\ 3(2x - 7) = 4(x + 9)$

(3 marks)

$ii)\ \dfrac{3x + 1}{2} = x - 1$

(3 marks)
(total 6 marks)

5. A circle has radius 6.5cm.
i) Calculate the circumference of the circle.

(3 marks)

ii) Calculate the area of the circle.

(3 marks)
(total 6 marks)

6. Work out the volume of the cube shown.

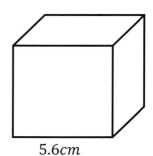

5.6cm

(4 marks)

7. $y = 4 - 3x$

i) complete the table below

| $x$ | $-3$ | $-2$ | $-1$ | 0 | 1 | 2 | 3 |
|---|---|---|---|---|---|---|---|
| $y$ | | | | | | | |

(3 marks)

ii) Plot the graph of $y = 4 - 3x$ on the grid below.
(clearly label your $x$ & $y$ axes.)

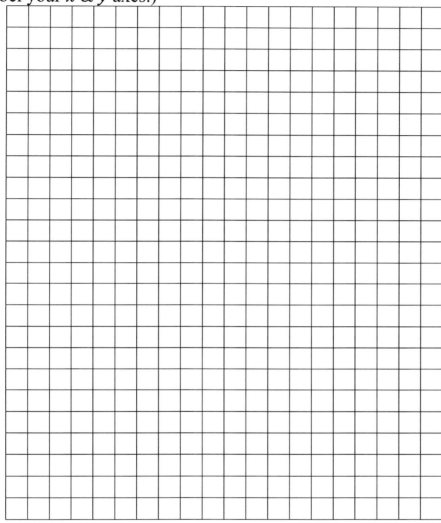

(3 marks)
(total 6 marks)

8. Work out

$$4\frac{2}{3} \div 2\frac{1}{3}$$

(5 marks)

9. Round the following numbers to 3 significant figures.
i) 2015

(2 marks)

ii) 2.021

(2 marks)

iii) 403.5

(2 marks)

iv) 1.01256

(2 marks)
(total 8 marks)

10. Work out the area of following shapes.
i)

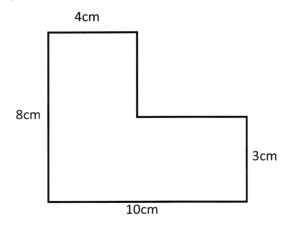

(5 marks)

ii)

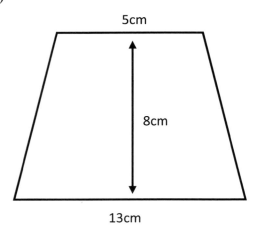

(5 marks)
(total 10marks)

11. A sequence has terms

$$12, \quad 19, \quad 26, \ldots \ldots$$

i) Find an expression for the $n^{th}$ term.

(4 marks)

ii) Find $25^{th}$ term.

(3 marks)

iii) Is 100 a term in this sequence?

(3 marks)
(total 10 marks)

12. Work out the following.

$i)$ $\sqrt{225}$

(2 marks)

$ii)$ $7^3$

(2 marks)

$iii)$ $4^3 - 2^5$

(4 marks)
(total 8 marks)

13. Factorise fully

$i)$ $4a - 10 =$

(2 marks)

$ii)$ $24y + 18 =$

(2 marks)

$iii)$ $2x^2 + 20x =$

(3 marks)

$iv)$ $24y^2 - 32y =$

(3 marks)
(total 10marks)

14. Table show goals scored by a football team in matches.

| Number of goals(x) | Number of matches(f) | |
|---|---|---|
| 0 | 7 | |
| 1 | 9 | |
| 2 | 8 | |
| 3 | 4 | |
| 4 | 3 | |
| 5 | 1 | |

Work out the average number of goals per match by this team.

(5 marks)

15. Calculate the area of triangle drawn below.

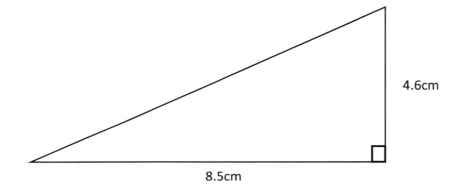

4.6cm

8.5cm

(3 marks)

Total for paper: 100 marks

**End**

# Answers

| Paper A1 | Paper A2 |
|---|---|
| 1. i) $\frac{2}{3}$, ii) $\frac{4}{9}$ | 1. £6 |
| 2. i) 22, ii) $\frac{2}{5}$ | 2. 0.0210, 0.1002, 0.1021, 0.2010, 0.2101, 0.2111 |
| 3. 4 km/h | 3. i) $18.85cm$, ii) $28.27cm^2$ |
| 4. i) $5x + 19$, ii) $x^2 + 5x + 6$ | 4. i) £45, ii) £90 |
| 5. i) $3n - 1$, ii) 149 | 5. i) $2 \times 2 \times 5 \times 5$, ii) $2 \times 2 \times 2 \times 3 \times 5$, iii) $2 \times 2 \times 3 \times 5 \times 5$ |
| 6. i) $x = 8$, ii) $x = 7$ | 6. i) 4.2cm, ii) 21.4cm |
| 7. i) $\frac{3}{4}$, ii) $\frac{7}{25}$, iii) $\frac{7}{20}$, iv) $\frac{12}{25}$ | 7. $a = 28°$ (alternate angles), ii) $b = 28°$ (corresponding angles), iii) $c = 152°$ (allied angles) |
| 8. John 9 sweets, Mary 15 sweets. | 8. 13.84 |
| 9. i) 8, ii) 49, iii) 512, iv) 256 | 9. $\frac{11}{12}$ |
| 10. correct diagram must be shown | 10. i) £1,875, ii) £10,625 |
| 11. i) $10,000g$, ii) $1.5l$ | 11. i) 25, ii) 31, iii) 6 |
| 12. i) $x = 60°$, ii) $x = 30°$, ii) $x = 35°$ | 12. i) 1750m, ii) 7.65m, iii) 250m, iv) 0.82m |
| 13. i) $3(x + 5)$, ii) $2(2y - 5)$, iii) $x(x + 7)$, iv) $y(y - 8)$, v) $3x(x + 4)$, vi) $5y(2y + 5)$ | 13. correct net diagram must be drawn. |
| 14. i) $HCF = 16, LCM = 192$, ii) $HCF = 18, LCM = 108$ | 14. i) |

14. i)

| $x$ | -2 | -1 | 0 | 1 | 2 | 3 |
|---|---|---|---|---|---|---|
| $y$ | -5 | -2 | 1 | 4 | 7 | 10 |

ii) correct plot must be shown

15. i) 10, ii) -14

| Paper B1 | Paper B2 |
|---|---|
| 1. 10 | 1. £5.46 |
| 2. i) 6, ii) 2 | 2. $i) HCF = 16, LCM = 240,$ $ii) HCF = 18, LCM = 360$ |
| 3. i) 3.22, ii) 23.072 | 3. 100km |
| 4. 35 | 4. $i) x = \frac{57}{2}$ or $28\frac{1}{2}$, $ii) x = -3$ |
| 5. i) $34cm^2$, ii) $25cm$ | 5. $i) 40.84cm, ii) 132.73cm^2$ |
| 6. $P = 5x + 2y$ | 6. $175.616cm^3$ |
| 7. 6.4, 7.1 & 8.5 | 7. i) |
| 8. 2cm | |
| 9. i) 125, ii) 125 | |
| 10. $x = 80°$ (allied angles), $ii) y = 100°$ (opposite angles in a parallelogram), $iii) z = 80°$ (allied angles) | ii) correct plot must be shown. |
| 11. i) 12.5, ii) 43.2 | 8. 2 |
| 12. i) $x^7$, ii) $x^5$, iii) $x^3$, iv) $x^7$ | 9. $i) 2020, ii) 2.02, iii) 404, iv) 1.01$ |
| 13. $a = 79°$, $b = 101°$ | 10. $i) 50cm^2, ii) 72cm^2$ |
| 14. i) $x = 4$, ii) $a = 2$ | 11. $i) 7n + 5, ii) 180, iii) No$ |
| 15. $2 \times 3 \times 7$ | 12. $i) 15, ii) 343, iii) 32$ |
| | 13. $i) 2(2a - 5), ii) 6(4y + 3), iii) 2x(x + 10),$ $iv) 8y(3y - 4)$ |
| | 14. 1.6875 |
| | 15. $19.55cm^2$ |

For Paper B2, question 7 i):

| $x$ | -3 | -2 | -1 | 0 | 1 | 2 | 3 |
|---|---|---|---|---|---|---|---|
| $y$ | 3 | 10 | 7 | 4 | 1 | -2 | -5 |

Printed in Great Britain
by Amazon

80076365R00031